Diary of a Small Farmer

by
Susan G. Mehaffey

AuthorHouse™
1663 Liberty Drive, Suite 200
Bloomington, IN 47403
www.authorhouse.com
Phone: 1-800-839-8640

AuthorHouse™ UK Ltd.
500 Avebury Boulevard
Central Milton Keynes, MK9 2BE
www.authorhouse.co.uk
Phone: 08001974150

© 2006 Susan G. Mehaffey. All rights reserved.

No part of this book may be reproduced, stored in a retrieval system, or transmitted by any means without the written permission of the author.

First published by AuthorHouse 9/13/2006

ISBN: 1-4208-9472-2 (sc)

Printed in the United States of America
Bloomington, Indiana

This book is printed on acid-free paper.

Bloomington, IN  Milton Keynes, UK

Dedication

Dedicated to Jacquie, Kyle, Alexis, Allie and their grandparents… none of this would have been possible without the latter's undying love, energy, and financial contributions. And in loving memory of my cousin, Beth Parrish, who committed suicide related to unemployment, alcoholism, and being tired of trying. Also in honor of Hurricane Katrina's victims and young and old veterans and their need for jobs.

Table of Contents

Introduction		vii
Chapter One	Free Range Rabbit Husbandry	1
Chapter Two	Swine Husbandry	15
Chapter Three	Chicken Husbandry	19
Chapter Four	Dairy Calf Husbandry	23
Chapter Five	Organic Horticulture	27
Photos		30
Epilogue		47

Introduction

Hello…my name is Susan and I have some interesting facts to share with you if you like animals, horticulture, or new concepts and ideas. I am a retired R.N. experienced in hospital and geriatric center patient care with some supervisory experience. After four years of education and nine years of experience, I found myself unemployed when I requested off the night shift. Unfortunately, all in life is not fair, but you may be surprised to find how fortunate termination for insufficient reasons may be. I had just purchased four and a half acres intended for raising a couple of horses and a young child. Because of family contributions, my family and I set up a small farm and I have raised hundreds of various farm animals at one time or another during the 1990's and several pets. My daughter has been in 4H and continues to thoroughly enjoy and appreciate all animals and truly understands what it takes to have food. Her father was a commercial crabber at one time and she sometimes assisted him as well.

I have kept journals of animals and horticulture over nine years and have summarized here in case some of you readers would like to learn through someone else's trials and tribulations. Farming is an expensive venture and though I hesitate to say it, it does not seem profitable, no matter how long you work at it. But do you choose to make a difference in the children's future or blame everyone else if there are no valuable food resources available? Prevention is the best medicine. Let us work toward preventing malnutrition and hunger as undeveloped land becomes scarcer, and aid those countries that are dealing with this plight, while providing employment for our own and theirs. How can we do this for and on the path of least resistance? Read on and be prepared for a shocking revelation! Here is a clue… perhaps you may become a part of the great food chain indirectly one day.

Having been an active-duty nurse, I realize that I came prepared for such things as injections, tube feedings, lancing, wound irrigation, birthing, and assessments. How do you learn if you have never done these things? If you are fortunate enough to afford a veterinarian at anytime, they will probably be very happy to teach you during the procedure on your animal; otherwise, perhaps this book can serve as a text for the basics. Animal science majors and their educational material would be an excellent resource. There are also quite a few good ideas in this book that may spark your imagination and ingenuity. Some of you may think the ideas crazy! Regardless, I hope that the reading is entertaining and informative. And I wish to promote further agriculture engineering and research that I pioneered.

    I would like to acknowledge the assistance of dedicated small and large farmers whose stock I purchased. Without their work, I could not have accomplished mine. They loved their animals and took good care of them. And, of course, who can do without a good veterinarian group? Ours is called Country Oaks Veterinarian Clinic. I wish to thank a fellow farming enthusiast, Tony. He was a hard-working blue-collar worker whom managed to fit in hog husbandry with me in the mid-90's, sometimes with black circles under his eyes, and later owned his own hog farm. And Bryan and Donna were willing to swap piglets for welding work. I must thank my parents, sister, brother, and ex-husband for their tireless labor. It would have been much harder and slower was it not for the countless dinners, babysitting, fencing labor, handyman repairs, and the occasional front-end loader and dump truck work. Thank you leaf rakers…countless bags were collected over the years from suburban driveways for vegetable gardening and the miraculous vine and tree pits described in chapter five. Thanks to Winn Dixie and the continuous supply of garbage refuse and to our public school system that allowed swill pickup. Also I would like to thank the Writer's Ocala Workshop; we offered praise and constructive criticism to one another about our writings. Our old neighbors, Bob and Brad, and my twin sister, Bonnie, have been helpful in the same way. I wish to give thanks to Mr. John Dieterely, a former regional director of Heifer Project International, whom Tony and I met in 1994. He encouraged volunteering for his outfit by just doing what we were already accomplishing. And I could not have xeroxed as many times and as affordably as I did without the use of our local Mail Boxes Etc. services, now UPS. And what a cheap way to have your own office! Thank you to my professional artist neighbor, Irene Myers, whom has owned her own business for eighteen years. And my daughter's typing and computer skills sure were needed. Last but not least, thank you to my publisher, AuthorHouse. I have revised more times than I can count and I appreciate their patience and affordability.

    I now no longer own and operate the small farm that our family built, but was pleased to have sold it to other farming enthusiasts. The opportunity to own and live in a house close to my aging folks arose, but I do miss the animals some. They can provide you with a sense of satisfaction at a job well done and worthwhile. And thank you for embracing all the ideas that will unfold to you as you read this book. Maybe you will think of more futuristic ideas that can provide many more goods and services. (Unemployment can be brutal but a godsend if affordable and one has good goals). Good luck at your own ventures.

Chapter One

Free-Range Rabbit Husbandry

The Elusive Goldilocks was a rabbit. It was so named because it was not possible to catch it up in traps, though I only tried a few times. Was it a buck (male) or a doe (female)? Sometimes I could accurately guess the gender of some of the rabbits by their behavior towards one another in our fenced free-range pasture with three man-made underground warrens. Our fence actually had large enough rungs on one side that allowed the mammals to come and go freely into surrounding fields, but prevented dog invasions into their pasture. Occasionally, our dog, Squeaky, was quicker than the unsuspecting rabbit or bunny, though. Refer to the photos displaying some of the rabbits in their man-made, free-range habitat with the covered homemade warrens. That will help you visualize what I am talking about; the construction is described later in this chapter.

Meet Thin Stripe, Emelius, California Gold, Brown Spots, Grey Beauty, Oreo I and II, Seminole, White Paw, The Wild Thing, Big Mama and numerous others. Perhaps bunnies living like this are destined to be your Easter present or dinner. You may even wish to provide housing for the Easter Bunny mascot! Some of these critters averted owl, falcon, hawk, and dog attacks, and when nearly full grown, were ready to breed and reproduce when they chose and with which mates. What an interesting and educational experiment, though they all did succumb to predators in the long run. The freedom provided utopia for the rabbits and our fowl too. I did have to separate the two due to a problem mentioned later. I am remembering the words of a great blues singer, Louis Armstrong…"I think to myself, what a wonderful world."

Susan G. Mehaffey

My daughter and I started the free-range rabbit project in our back pasture in the winter of 1992. It began as a safe haven for the pet rabbits from our neighboring German shepherds who actually broke through a shed I used for our first two rabbits. There were a total of seven litters born inside the "Funny Bunny and Fowl Corral" pasture between February 1993 and May 1997. The number of kittens per litter ranged between seven and ten; and these few does were only five to seven pounds. The rabbits were mixed breeds. I wish that we could boast about more colonies, but I can only pass on education. I could not afford to be more aggressive about introducing new breeders due to limited funds and various predators. Most of these vegetarians were suspected to have been eaten by birds of prey, as evidenced by a tuff of rabbit fur in their pasture now and again. Then overhead netting was installed, but it appeared that neighboring cats or foxes tore through it in several places. Once, I did take a break from raising the free-range rabbits for several months at a time, but the wildlife predators apparently have good memories and returned.

One of the lessons learned with this agricultural engineering project was that interbreeding does occur. How else did the earth become so populated with wildlife? I do believe that sisters and brothers and half-siblings do mate and with multiple partners each breeding cycle, as evidenced by the various colors and markings in one litter. I do not know for sure whether or not the parents and offspring will mate.

I did trap a one-year-old buck that was apparently castrated by the larger brother or half-brother, or both. They were the only two rabbits left besides the Elusive Goldilocks by May of 1997. The victim's ears were torn, tail fur missing, and testicles nibbled off from the cord. Perhaps the Elusive Goldilocks had been involved. A local rabbit farmer said that when two bucks are placed in a cage together, the larger will chew the testicles right off the other! Remember, our setup allowed the rabbits to leave for open pastures and briar patches anytime they wanted. Free-range dogs and wildlife, such as red fox and opossum, live around the pastures. Was the younger buck an unsuspecting victim of this own castration by his fellow bucks, or aware of his fate and opted for safety in numbers inside their fenced-in pasture, versus being a dominant buck of his own isolated domain outside that pasture with the man-made dens? If this peer fate is too cruel for you and your comrades, consider taking your juvenile buck to your veterinarian for castration or using rubbing alcohol in a towel to anesthetize your rabbit. I am a hearty old gal and castrated piglets with an ice block and scalpel, but I chose nature to take its course concerning the rabbits. Each farm manager will have to decide breeding choices. Cages are a choice necessary if selective pairing is desired because does will breed with more than one buck in a free-range habitat. Hence, the reason for half-brothers and half-sisters in the womb at the same time.

The University of Florida offered a one-time rabbit class open to the public. Animal science, horticulture, and other agriculture majors turn into 4H and FFA leaders and agriculture extension agents, who can be found at your local county extension office. Unfortunately, they will not know much about free-range rabbit husbandry. I am a pioneer and wish to teach anyone who may be interested. And it is best to learn from my mistakes if you wish to have productive farms with less trial and error. Perhaps these kinds of habitats will be established as large farms or parks around the world one day. Perhaps it is even possible to use a similar model on rooftops! You certainly would not have the problem of preying foxes, opossums, raccoons, snakes, and cats digging under, crawling through, or over the fence!

Before we continue learning more about free-range rabbits, I would like to introduce myself as a member of an international group of volunteers devoted to alleviating hunger and malnutrition forever: Heifer Project International (1-800-422-0755). I believe that all of us have the shared responsibility to extend this goal to all farm and domesticated animals too.

The loveliest aspect of this new rabbit-farming technique is that the main staple is grass and that the rabbits were free to roam, while fighting hunger at the same time. Their weight gain will not be as impressive if day-old bread or commercialized feed is not provided. And the rabbits loved the slightly aging fruit and vegetables that grocery stores threw out, which provide a lot of vitamins and minerals. Some of the rabbits have been observed hopping toward me when I had day-old bread in hand. Just imagine what a musician and his or her instrument could do in regards to training of these animals …another Pied Piper? Sometimes we were allowed to pet them when they were first released; it did not take them long to become elusive, though. One of the most interesting behaviors to witness is the jumping in the air and sometimes a sort of "leapfrog" game. I also enjoyed watching them sit up on their hindquarters and look around; maybe the rabbits were sniffing the air too. Caged rabbits do not have this option.

Another exciting feature about the free-range environment is that it is relatively inexpensive and easy to set up a pasture of man-made warrens. This may provide much-needed relief for victims of kwashiorkor or marasmus (protein malnutrition) in third-world countries. Plus the man-made warrens offer shelter against hurricanes, tornadoes, and intense heat. Of course, earthquakes and flood regions may be fatally dangerous for these fair creatures. So, here is how I made the homemade rabbit warrens: it took a front-end loader to begin construction of the warrens, concrete blocks for lining the edges of the dug-out pits, a post hole digger to facilitate multiple nesting in each warren, wooden or metal posts to act as roof trusses over the rabbit lounges held firmly by the concrete block holes, and heavy welded wire on top of the posts to prevent dog attacks to each underground den. Then mounds of dirt that first were excavated from the pits were piled on top of the welded wire. And small sections of PVC pipe were used for openings into the lounge. Then add grass seeds, and voila...the start of something big: the food chain. I actually added another warren on top of the covered pits by cutting fifty-gallon plastic drums in half and covering them with soil and grass seed, or sunflower seeds. The latter only take about ten days to sprout, and the rabbits relished these! The top domes are where the chickens liked to lay. Try putting your hand inside one of them; it feels like built-in air-conditioning! Thank goodness for that because the sunshine in Florida only allows the rabbits to come out of their homes in the early AM and late PM hours, except for in the winter, when it is cool. And though overcast skies do occur from time to time, it did not change the rabbits' biological clocks.

We have noticed that the rabbits survive well regardless of whether water was available in or near the free-range pasture. Apparently, early-morning dew is enough. This can be a great help in missionary work, where water is scare. But I recommend having a clean source of water available at all times, particularly when does are nursing their kits. If raising ducks among the rabbits, line water bottle feeders along the fence; ducks and ducklings preen in even the smallest containers of water and poop in them too! If not cleaned out daily, the chickens may succumb to fowl cholera. Oddly, the rabbits probably drank from a man-made small pond that was well-used by the ducks, but never once became ill. It is probably a good idea to pour in an anthelmintic into the water supply, such as a feed store five-gallon upside-down bucket with a trough. Perhaps the anthelmintic can be added to their commercial feed one day. This drug kills intestinal parasites.

These are the four most important things to remember when raising free-range rabbits: are they healthy, breeding, reaching their maximum weight potential, and safe from predators and disease? Occasional trapping is needed in order to check weights and for inspection. I learned through observation that the moms settled into one homemade warren or another for several weeks; hence, a baited trap placed at the opening of her warren may trap her and the offspring together. There are small traps designed perfectly to capture just the weanlings. Pick the large rabbits up by their ears and grab a chunk of back fur. They do have sharp nails and will scratch and kick themselves free if not held securely. Thus far, the cutie-pie weanlings weighed between six and ten ounces when they first emerged out of their underground warrens. I imagined that they were about three weeks old, based on the weight of caged bunnies. The first one that was ever witnessed was six ounces and did not know what a human was; it simply allowed the children to walk up to it and "pluck" it. We played with it for a while, then released it, and it went back to its siblings and mom, never to allow us to catch it like that again. That particular bunny had a beautiful all-white coat with black eyes and black lining around the eyes. It was rare to find two rabbits with the exact same markings, so recording data on them was easy.

I would like to point out six problems you might encounter with free-range rabbitry. Here is the first one; be prepared for a foul-smelling discharge upon draining. Pasteurella abscesses are rare, but occasionally a rabbit will have a pocket, or pockets, of pus that can grow to enormous sizes. They can be treated by your local veterinarian, or yourself, with draining and antibiotics. The children and I used rubbing alcohol in a towel to put one infected rabbit to sleep temporarily. Farmers may sometimes cull them; removing them so they do not cross-contaminate. Do spider bites cause these or a sexually transmitted microorganism? The U of F rabbit class that I took suggested that it might be the latter. The animal observer, or ethologist, plays an important part of history. I have only had two cases of Pasteurella and the rabbits were related. The grandfather had died long before the grandchild was born. I believe that spider bites are the real culprit. The grandchild had small abscesses on his trunk which could not be seen by simple observation; hence, trapping is necessary.

Fleas can be a big problem, in particular, stick-tight chicken fleas. Apply dog and cat flea spray p.r.n. (as needed) ; the fleas tend to latch onto the rim of the ears and around the nose, eyes, and mouth. Oreo II was so grossly infested that he drowned while I was attempting to wash and de-flea him; he was so anemic that he simply fainted. The stick-tight chicken flea infestation arose from chickens that were allowed to live in the upper warrens and turn them into nests. (They laid so many eggs that the eggs could not fit under the rotating hens and some were repeatedly rolled out by them. I spotted one hen eating an egg, unfertile and half-scrambled, by their own incubation! The automatic incubator is the only solution to full productivity.) I obtained the chickens and the used automatic incubator from a family that raised chained dogs near the chickens, and a simple inspection of the birds before release into a free-range pasture can prevent the stick-tight chicken flea problem. I learned this one by trial and error and found that two or three applications of lime and Sevin (flea dust) in the chicken bedding and direct spraying mentioned earlier did rid us of the infestation. I only came across one or two fleas on another rabbit or two after those treatments.

Here is an interesting tidbit related to the flea problem; Brown Spots was relocated to a briar patch about eight acres away from his warrens after running into town with him for flea spray and dropping him off after the treatment and before returning home. The next morning, I found him just outside the free-range rabbit pasture! How did he figure out that one? Can animals read one's brainwaves? I think so, based on other thoughts and the dogs' and rabbits' responses. Of course, Brown Spots may have smelled the warrens.

We only had one case of malocclusion of the two bottom front teeth. They grew so long and crooked that their length could easily be seen as Oreo I meandered around the yard far away from us. Pruning shears resolved that problem.

Free-range rabbit husbandry may be new on the horizon and it may take years to discover which management techniques yield the most meat and fur. This kind of free-range rabbit farming can be so easy; a dream for anyone who is on their feet a lot! The next problem to watch for while inspecting the rabbits may make you cringe and you may have to evaluate your true hardiness for handling this problem that may arise occasionally. If it is not noticed readily, perhaps it can be a gruesome and prolonged agony for the doe. There are grub worms in the soil that will crawl into a doe's vagina. Our veterinarian figured out what was wrong with the first one affected and extracted the partially exposed worm from her strange-looking posterior. She was treated with penicillin IM (intramuscular) and the inflammation/infection healed rapidly. Then she gave birth to two sets of beautiful offspring before succumbing to a predator. Her name was Seminole and she had two litters within three months; whereas the other doe, White Paw, had two litters within two months. White Paw was a surviving offspring of Big Mama, and both Seminole and she were approximately ten months old at breeding. For some unknown reason, Seminole was feasting on grass in the rain and allowed me to bump right into her in the night and pick her up. I realize that she was a mom when I felt her nipples. She weighed five pounds. Two does got into a fight in a cage and then were released into their free-range field that same day. Apparently, one had two fur tears and two grub worms crawled under the fur and had to be extracted the next day after I coincidently trapped her.

What do the weights tell us about productivity? On one hand, the rabbits have a better quality of life than the caged ones, but increased activity burns off calories. And how many calories are gained from grasses anyway? Improvements have been made in weight gain since removal of the free range chickens, since hens have fairly hearty appetites and eat a lot of day-old breads and garbage refuse intended for the rabbits. An eleven-ounce weanling re-released in January 1995 weighed in over two pounds heavier in just three months after the herd of free-range chickens was relocated on the farm. Previously, it had taken a year for a castrated buck, Cosmic Creepers, to reach six and a half pounds. I wish that I could give comparisons with caged rabbits of the same breeds, but I am merely an amateur animal science researcher. I suspect that university agriculture professors have that information, or would obtain it if legislation and funding permit. Our local agriculture extension office did supply me with this bit of information: caged doe and litter will consume approximately fifteen pounds of pelleted feed per young, weaned rabbit. In case you are alarmed by the low weights mentioned above, keep in mind in that these were not the large white New Zealand breed.

It is my wish to advocate a free-range rabbit habitat with cages, the cages being for isolation, selective breeding, nesting, or inspection. Imagine walls of horizontal pipes that have cages attached to them on the inside of a gymnasium-sized barn. Then imagine perpendicular pipes attached to these horizontal pipes and they travel through the wall and onto a grassy hill. The outside PVC pipes could have different-colored rings to help the rabbits identify the opening closest to their own cage and nest if in fact, they really are not color blind. Imagine animal science majors and their helpers operating a man-lift to upper cages. The equipment should be caged to protect the rabbit from falling from great heights during inspection and weighing.

Imagine what the outside of the barn would look like. The hill, starting from the outside wall, would lead down to a beautiful garden, perhaps, and the field would be covered with a wire mesh overhead to protect against predators. Or the hill would start in the center of the field and allow for trap doors on the top of the PVC pipes to better facilitate proper cleaning of the interior of the PVC pipes. As I already emphasized, this type of environment obviously facilitates more room to move about, sit up, run, and hop. Be careful to provide ample and large enough nest boxes if there is breeding stock. What a lovely farm that would be for both worker and animal. Please refer to the described drawing by a professional artist and the three-dimensional model photo. The 3D model picture demonstrates two versions of the hill. As I mentioned, it may be best for sanitation purposes to go with the PVC pipes that have doors and extractable grids inside to prevent stomping on their excrement.

Now let us discuss the sixth problem and the most important one to consider. All of the pasture companions were killed by wildlife or dogs before their productivity, weight, and mating preferences were fully realized. The neighboring dogs, plus our own, were medium and large and one time the neighboring German shepherds did gain access into the rabbit pasture and demolished one of the warrens. It is demonstrated in the photo section. Fortunately, our rabbits dug tunnels throughout their man-made warrens, and this did save some of them. But I did witness that the neighboring dogs killed for sport and left carcasses to be eaten by buzzards. A dominant buck, Oreo II, did build a deep hole in a sandy cliff around a man-made pond that receded constantly. He was attempting to build more of them before he died. I suspect the rabbit that occupied this hole ran for cover in the man-made warrens. Please learn from our errors. Tall fencing is needed with a strand of barbed wire on top. Perhaps hot wire on the bottom would alleviate the need for concrete blocks. Or better yet, have the fence dropped into the ground after heavy equipment makes a trench. We utilized heavy blocks from demolition sites and set apart just enough to fall into a digging culprit. A short fence was improved with decorative trellis wiring on one side, but alas, the neighboring German shepherds proved how crafty they can be by squeezing through and bending back the wire until an electric strand shocked their memories! No more dog attacks then. Foxes in the area may have had lush pickings unless the roaming dogs scared them off. I have never trapped one, just opossums. That is where we got the picture of the baby opossum. The rabbits that ventured outside of the fences repeatedly demonstrated an excellent memory of their escape routes; the fences were not designed originally with the rabbits in mind. They only had a few openings in between the concrete blocks to the lowest fence rungs and the rabbits did venture into the neighboring pasture occasionally. I frequently witnessed our faithful dog companion, Squeaky, chasing them back into their pasture with the three homemade warrens. Squeaky's intentions were to eat them.

Despite predators eventually consuming our entire crop, I still advocate free-range rabbit husbandry; but obviously, not just like ours. And of course, some may be interested in farming snakes, iguanas, turtles, rodents, etc. To be honest, common sense tells you that urine and feces may be accumulating inside the warrens and perhaps even a decomposing rabbit. I did smell excrement when up close to an opening into the warrens. Professional engineering would certainly be needed to improve upon this primitive backyard sanctuary. For those more serious about my version of a more efficient and commercial free-range rabbitry, consider some more suggestions. The horizontal PVC pipes attached to the cages and the barn walls ought to be slightly slanted with large rectangular openings on the surface of the pipes to allow for easy withdrawal of grids and hardy cleaning. A garden-hose or pressure-washer could wash the excrement down into a gutter from the inside PVC pipes, and be deposited in the meadow. Then the waste could be recycled. And perhaps there should be weather strips on the end of the PVC pipes to reduce the heat and air-conditioning bills inside the gymnasium-sized room. Here is an intriguing idea…a restaurant where the customers dine while watching free-range rabbits and bunnies graze, and actually see inside a warren or two! It makes me think of vineyards in California where visitors taste samples of wine and hear how it is made. Perhaps community science centers will build some exhibits for schoolchildren, where a clear-view window has a door, and the children can touch the bunnies and rabbits. Imagine the possibilities! I sure wish that I would win a big lottery one day soon! What fun I would have with this new farm! I also imagine attaching human shelter for senior citizen day cares, hospices, or severely handicapped children's schools, since I am a caregiver also interested in quality of life once permanent disability sets in and all one can do is watch. My mother has severe Dementia and enjoys looking out on her gardens and pool from her kitchen table. There is not much activity going on except for flying butterflies, though. There is another advantage to a free-range rabbit habitat versus a caged habitat. I have seen sores on the rabbits' feet due to prolonged pressure on the wire. What relief a free-range rabbit farm would bring to caged animals in particularly small quarters.

I am no professional engineer but I think some even more fine details of this new kind of rabbitry need to be considered. In case anyone is actually contemplating building this ideal commercial farm, please keep these things in mind too:

1) Tight-fitting caps for the outside PVC pipes in order to prevent flood waters from coming through the walls need to be considered. Flood gates inside the walls for double protection also need to be considered.

2) Automatic locking doors exiting into the first PVC pipes from the cages are needed, in order to keep the rabbits caged if you desire to trap them for weighing, inspecting, harvesting, and so on. Perhaps there is a more economical solution other than automatic locking doors. But I have not figured it out yet, unless mere trapper cages are utilized.

3) Cages may need to be able to be pulled away from the attached PVC pipes in order to scrub the feces of the bottom of the cages, particularly if laden, and for other purposes, such as for the transport of the animals.

4) Poop trays under the cages to catch the feces are necessary. And let us not forget to recycle!

5) Remember that compartmentalization of the rabbits may be necessary. The blue ink stamp rabbits in the three-dimensional model photo symbolize the weanlings and juveniles that are males. One wall of cages has stickers representing the breeders and their nest boxes, plus there are pink ink stamp does or juvenile females in the field with a few bucks. In other words, it may be best to have one field of juveniles and at least one field of breeding stock.

6) Outside fences starting from the corners of the building are necessary to separate the fields of the colonies mentioned above.

7) Reminder of what was previously suggested: all inside PVC pipes need to be slanted to allow proper cleaning and drainage into the meadow by way of gutters located at the corners of the building. This one is not demonstrated in the 3D model.

8) Reminder: rings of different colors around the PVC pipe openings in the meadows may be necessary to help the moms best identify which PVC pipe is closest to their cage and nest box. A color ring on each individual inside cage may be a good idea too. This detail is not demonstrated by Irene in her drawing. This could provide further research about color blindness of the animals.

9) Reminder: the outside PVC pipe model in which the hill starts in the middle of the meadow is a more sanitary solution, allowing access to the extractable grids inside the exiting pipes.

10) Reminder: weather strips attached to the outside PVC openings may be a good idea to reduce heat and air-conditioning bills.

11) Reminder: the man-lift needs to have a cage built around the platform to prevent rabbits from injury.

Here is a summary of some commercial possibilities with free-range rabbit husbandry and the jobs it could produce. Can you think of any more?

1) City planners and parks and recreation managers could be educated about free-range rabbit husbandry and requested to consider city retention ponds and parks as sanctuaries for unwanted pets. Remember to have professional engineers improve upon my primitive example…please!

2) More trappers may be needed to retrieve the bucks and does that are ready for harvest or swapping if city retention ponds and parks are used as sanctuaries for the unwanted small animals, birds of prey, and sightseers. We need to evaluate whether or not is it a good idea to promote more birds of prey that would most likely follow. I do wonder if their excrement could be a disaster waiting to happen if it is contaminated somehow. The birds of prey do help control the rodent population, though.

3) More wildlife and endangered species officers may be needed to set traps and handle potential predators. We used to have at least one beautiful red fox in our area until killed by a vehicle. I do speculate that their only homes will be zoos one day and perhaps much sooner than most of us can imagine.

4) More Disney artists, zoo architects, and the like, may be needed to develop first-class exhibitions and/or farms. Service positions are then made available to serve the small animals and visitors amongst the man-made utopia.

5) More butchers may be needed to aid in the fight against malnutrition and hunger.

6) Perhaps there will be a dramatic increase in small farmers and hobbyists. Picture this if you will: a home with front and back French doors or large glass windows facing underground warrens, then brick walls, or fencing, in the background to prevent predator attacks. Overhead plastic construction netting or welded wire could be used to keep out the birds of prey, wildlife, dogs and cats. Driveways allow access to suspended cages

inside the courtyards which can be lowered into pickup trucks by one or two persons. 4H and FFA memberships on the rise!

7) The Peace Corps and Americorps may find it in their best interest to learn about free-range rabbit husbandry and force-fed organic fruit and nut tree production. This rapid method of plant growth is the subject of Chapter 5. It really is possible to have a tree nearly double its size in two years and have fruit on it just a few months after transplantation. Then there is more work that can be done in the second and third worlds were water is more scarce, as well as the first. Squirrel meat on the rise? My brother grew up squirrel hunting and loved this meat.

8) Easter Bunny parties and stories will interest the young and old. I remember when our local tourist attraction used to have a Santa Claus home where you would visit with him. How about the Easter Bunny home, where you visit him?!

Susan G. Mehaffey

Summary: The Start of Something Big

There once was a middle-aged woman,
Who had so many big, aggressive neighborhood dogs,
That she feared for the caged and wild rabbits,
That lived on her four-and-a-half-acre home site,
Among her daughter's dogs, cats, chickens, ducks, and hogs.
Plus she believed in a good quality of life for all animals.

So she devised a free-range pasture with man-made warrens and pond,
And had pits dug out by a front-end loader.
The pit sides were lined with concrete blocks,
And a post-hole digger encouraged multiple nests.
Then wooden posts were arranged to serve as the roof trusses for the mounds of dirt and grass above.

So picture this if you can,
A garden of grassy mounds and barricaded fence line.
The dogs could not plunder and destroy the warrens, nor scare the rabbits to death in their cages.
And the dominant buck dug more tunnels inside their homes and along the pond embankment.
Now the does, bucks, and kits roam happily in the open pasture to exercise and to dine.

This caregiver and once-4H mom called herself an ethologist,
And found that chickens also love this fenced free-range pasture too.
Then the hens took over the dome apartments on the top of the dug-out pits,
To beat the heat and to share nesting of eggs by the dozens.
They will roost indefinitely until the caretaker obtains an automatic incubator that yields better productivity.

The owner and creator found that all was not nirvana.
Stick-tight chicken fleas caused anemia and death in the dominant buck.

Diary of a Small Farmer

The nests under the plastic domes had to be treated with Sevin and lime dust.
Each rabbit had to be randomly and routinely trapped for inspection and weight.
Will your animal husbandry result in larger colonies of healthy wildlife or sickly ones?

Just when you think that you may have the solution,
To an unequaled quality of life and an ever increasing rabbit and/or chicken herd,
You may be surprised to find such things as castrated young bucks and vaginal worms.
Not much can be done for the former except to fatten them up for the plate, ready them for a pet, or allow wildlife to consume them.
But worm extraction and penicillin facilitated pregnancy for Seminole and natural birthing occurred.

The first litter began emerging from the bottom warrens during the winter of 1993 after I introduced the first two bucks a few weeks earlier.
There were eight to ten bunnies each litter, weighing as little as six ounces.
There were only two does and two bucks in the Funny Bunny Pasture then
We found that multiple partners with one doe are a reality.
You will be delightfully surprised at what each womb produces.

The Funny Bunny Pasture and Lucky Ducky Pond did help the rabbit and fowl colonies have fun here,
And should be definitely be considerd in areas that are subject to hurricanes, tornadoes, and blistering heat.
It is physically hard on caged, furry animals in the Deep South in the summers.
They panted heavily and limited their movement during the day when there was no breeze.

Netting had been installed and even a few large New Zealand rabbits released.
But they all were strong enough to chew through the nylon if it is thin.
Birds of prey or foxes apparently tore through the nets,
And one dog attack forced hot wiring the fence line.
Covered rooftop sanctuaries are probably the safest habitat without fencing.

I believe that I have covered all the problems and have adequately summarized,
Since the January 1992 startup of this project four acres away in the back pasture.

Susan G. Mehaffey

Remember, if mixing fowl and rabbit colonies is absolutely necessary,
Supplementary feedings of day-old breads or vegetable refuse will be needed,
Chickens and ducks are assertive and eat hearty, and rabbits cautiously graze.

Good luck and may Mother Nature and Father Time bless us all in every generation, even though we will probably overpopulate the earth by 2500 AD.

Chapter Two

Swine Husbandry

If your passion is piglets, these adorable and playful creatures will eat almost anything that is perishable. The same is true for the rest of the herd. I found that swine did not like bell peppers, onions, and citrus and only eat zucchini and squash if nothing else is available. Surprisingly hogs will crush whole crabs and eat every bite. Commercial feeds are expensive, and swine do eat more than you might imagine because they have incredible growth potential. We sometimes supplemented with garbage food from grocery stores; the manager prefers that you arrive before 9 AM each morning to the back entrance and park out of the way of the delivery trucks. First come, first served…and share with fellow farmers if the need arises. I must complain that there is still too much waste found in the garbage bins at times. Why is it not possible for farmers to have their own fenced-in food carts near the grocery stores' garbage bins and place a sign stating, "Take at own risk"? It is frustrating to see so much waste and not being able to reach it when it is at the bottom of the garbage bin! I use to get inside of them sometimes and it was tolerated by this store's manager most of the time. Because the boxes and bags of aging fruit, vegetables, meat, and bakery products were so heavy, I found it difficult to utilize this resource regularly. I would get too tired after several consecutive days of pickup. I worked by myself. And though one morning's supply could last a few meals worth; it would take two to two and a half hours to go through the food, separate from the trash, put it into separate bins, and then clean up. I did not have trash pickup at my homesite and had to take it to a landfill nearby each day.

There is always swill or slop, but sometimes it is difficult to get restaurants to participate. We used a high school cafeteria's leftovers, and the dieticians were happy to supply to me, provided I showed a permit first. They simply scraped off leftovers into a plastic bag inside a large garbage can and placed it outside the cafeteria in a screened shelter to prevent animal invasions. And does swill go far over day-old bread. My daughter and I had a two-year-old that weighed in at 727 pounds, and when she was purchased she weighed fifteen pounds! And you will be happy to find that if raising swine in a pen, they are always very clean, and defecate only in corners. We used our pens for the new mothers, and pastured the older swine. The breeding boar(s) usually lived among the pastured females and daily observation indicated their estrus cycles every 21 days.

The swill permit is issued by the Department of Agriculture. A Department employee will visit your grounds and inspect for safety and legitimacy of your request. You must cook the swill for at least thirty minutes at boiling point to kill potential harmful germs. I found that it takes one and a half hours to cook a two-gallon pot of slop. Then it has to cool down before serving it. Hogs relish it as much as piglets and pigs do, but if you find that these large swine are walking on their knees to the food, they most likely have developed gouty arthritis due to too much protein in their swill. Just pull out the chunks of meat and feed them to your dogs, cats, and chickens, and you will soon see the swine up on their hooves again. We also switched to grocery store refuse for one of my gouty hogs; she could not get up at all. Mostly fruits and vegetables come from that source. It definitely worked…she sprang up and ran when the butcher carrying his gun came towards her. I shared this butchered meat with a friend of ours, who ate a lot of it over several weeks. He developed kidney stones!

Dear doctors at university animal science departments…I compared the electrolytes of two of our swill-fed hogs versus the University of Florida's grain-fed hogs, and found that our LDH levels were extremely high as compared with the U. of F. animals. What does this mean? The butcher did report that my butchered boar's liver did have cysts.

Birthing piglets can be hectic. Our hogs have never been in a farrowing crate, but I have seen the need for it, such as hogs accidentally lying on piglets and crushing them to death. A farrowing crate may not help some newborns, though. That 727-pound hog deliberately killed her two litters with her teeth, even when one got close to her face and had not touched her nipples with its sharp teeth. I do have a lot of success stories, but it means that you always try to diligently watch over Mom during delivery and keep the babies safe and warm. I would use this time to clip the newborns' sharp teeth with a nail clipper and cut excessively long umbilical cords and apply betadine to the cord. Feed Mom after she delivers (116 days after mating), and if weak during labor. Otherwise, she may not be quick or strong enough to respond appropriately to a piglet under her while she attempts to rise or lie down. Sometimes, she needs to reposition for comfort; other times, she needs water, food, or to eliminate her wastes. Not all of the animals are exactly alike. I tried supplying water via the hose to one sow while lying down postpartum (after delivery); she responded by getting up. So as you can see, some sows would prefer to help themselves p.r.n. (as needed). If you have a small litter and suspect that some babies were not delivered, Blue Cohosh works wonderfully. It can be found at a health-food store. Women have been known to use it for abortion; hence, the old Indian term for this drug is papoose, or squaw root. It helped deliver dead babies from one of our sows. If a sow is bagging in her loins but has not delivered after her due date, be prepared with Blue Cohosh. I do not know if that sow mentioned above would be ok for further litter productions after the dead fetuses are removed. This was the mean Mom, so I took her to the market.

Feed needed to be increased while moms were nursing their young. By the time the large litter of piglets were weaned, I had increased to five buckets a day half full with day-old bread and swill, and we never experienced weight loss in the mom. I noted the piglets gained a pound a day. Needless to say, the piglets prefer swill over hard fruits, vegetables, and grain, and were always plump little adorable creatures with tails a-waggin'. Just imagine eight to eleven of these cutie pies roaming around your yard…they will not wander too far from Mom, and she cannot slip under the fence. Fortunately, they do not explore during the night when the dogs do the same. Watch out for your housedresses, ladies…piglets will tug at them with their teeth when demanding food, and the older they get, the more chances that your housedress will become ripped!

Here is an economic tidbit when feeding grain pellets to a growing pig. I tabulated each cut of meat, once butchered of a full-grown hog, based on Winn Dixie's advertised prices in December 1991. It cost me $24 to raise "Jeremy," when only calculating the cost of feed for him. Fair warning about raising swine: They have ferocious appetites and will eat chickens alive, their eggs and poop too. Perhaps this is a dangerous combination for the personal and commercial butcher and the sanitation of his facility. Penned swine have been known to eat small children and adults. The farm neighbor told me a story about an old man knocking himself out accidentally by the corner of the hog shed, and his hog ate him. A farrier told me about his four-year-old cousin being eaten. These pigs must have been starving creatures. Mine were not aggressive animals, though a boar and bull charged me once.

We did not use an electric fence, except for the gate to the rabbits, but I would advise it if you plan to raise boars among your herd. They will climb and bend down fencing, uproot posts with gates attached, bend up portions of your bottom fence line, and just plain demolish a good-looking fence. Your breeding program will be when they decide! They smell estrus cycles every twenty-one days when a female in not pregnant and will surely let you know.

I loved raising hogs, but garbage refuse feeding has got to be the most physically demanding farm chore to date. Be prepared for a lot of wear and tear on your vehicle too, because of the daily pickup of swill and/or garbage foods, plus your usual weekly errands. You will probably need a newer vehicle every six or seven years after starting off with a brand new one.

How about this commercial recycling idea? Swill collection centers in which the restaurant and school cafeteria leftovers are cooked and stored for present-day and future farmers. Just imagine the number of jobs this idea would create if available in every city throughout the world! One century or another from now, we will run out of cultivable land mass per populace and really have no choice but to save our slop. I theorize that it is the dairy cows that will have top billing for available grass and grain, and we will have to ask all non-farmers or non-gardeners to move into high rises. I predict that suburbia will turn into miniature rabbit farms on flat rooftops, and we will have to host commercial beehives among the suburban flower gardens. I did keep bees in the 1990's and rarely got stung. I wore street clothes only, but made use of the beekeeper's long gloves and smoker. One thing that stands out in my memory is the clear viewing window I had installed on one of the sides of the hives. It allowed my daughter to watch the bees work. I did place a curtain over the window for protection against the sun.

On the subject of resourcefulness, there are a couple of other thoughts. Perhaps pets and wildlife that become road kill will be given to crabbers by animal control to dispose of in creeks and rivers in order to feed crab and crawfish colonies, if this is an environmentally friendly idea. Maybe university marine departments will experiment with euthanized pet carcasses in order to study the short- and long-term effects on sea life and freshwater fish. There is more on seaburials in the Epilogue. I do predict that Central Florida will become the swill capital of the South because of the tourist industry.

Chapter Three

Chicken Husbandry

Maybe chickens, chicks, or eggs are more your style; keep in mind some helpful tips. Chickens require the egg maker grain, crushed shells, shellfish, or cheese in order to lay eggs. And if allowed to hatch their own eggs, they need individual nests provided. Otherwise, chickens and ducks will share nests, and too many eggs will invite the hen to push out excess ones. I have even seen a brooding hen eating one of the eggs, as I mentioned in the first chapter.

There were several broody hens that raised, or attempted to raise, their own chicks. I did have an automatic incubator that held up to ninety-nine eggs at one time. I usually had a new batch every two weeks. I would suggest collecting the eggs at least every couple of days and bringing them into the air conditioning to prevent a light-brown, oozing discharge – I think that I did incubate some very old eggs a few times. Perhaps this is why I experienced respiratory distress once after checking on my day-old chicks in the brooder chamber one night. To the best of my nursing knowledge, the epiglottis was stricken. I wonder if the rotten egg discharge may have caused contamination of the incubating chicks; I would suggest removing the rotten eggs immediately. Then I would suggest washing out the possible infected incubator with diluted bleach. The incubator manufacturing company sells an antibiotic spray. If at all possible, do not raise the incubating chicks and then the day-old ones in their heated brooder chamber inside your home unless you have proper outdoor ventilation. And while on the subject of health, always remember that lack of hand-washing is the number-one reason behind contamination. But this epiglotittis came from airborne germs. I once was babysitting someone else's parakeet while on the farm and experienced respiratory distress too, but the irregular breathing pattern was slightly different from the brooder chamber chicks' episode.

An automatic incubator is the best investment for consistent hatching of fowl eggs. The two inside pans need to be filled with water, so that drying out of the eggshells does not occur. Otherwise, the chicks will not be able to crack the eggs, and will die. Sprinkle duck eggs with water when checking your pan water levels every couple of days. It is ok to sprinkle the chick eggs too. The chick eggs require 99½ degrees Fahrenheit incubator temperature at all times. And duck eggs require 100 degrees, according to literature. But it is ok to keep it at 99½ degrees, and if it varies to 101 occasionally, it will not harm the developing embryos. I found that continual opening of the incubator to remove or check on the hatchlings may result in later hatches being delayed one or two days. This can be confusing if you have dried-out shells due to empty pans of water in the incubator, or too much calcium given to hens which produces excessively hard shells. The hatchlings then may not have enough strength to crack their shells. Label the eggs according to the hatch/twenty-first day. I have helped quite a few chicks hatch, but it is best to allow them to do it themselves because they may still be developing. Sometimes the hatchlings are born handicapped and these do not usually survive. It may be best to cull them so they need not suffer. A disabled chick will be eaten alive by ants if placed in the coop or yard.

A brooder chamber is needed to keep the new hatchlings warm. Being in Florida, they can be outdoors in only a few short days when the weather is warm. Chicks and pullets require coops that have covered yards, because of attacks from birds of prey and cats. Use the smallest chicken wiring around the bottom to prevent the critters from getting out of their yards and succumbing to dog attacks. Run chicken wire along the ground inside the pen to prevent dogs from digging under the pen. Or drop the fence line underground, as described in Chapter 1. You may have encounters from opossums, raccoons, foxes, and such. Free-range dogs and a pen near the house best prevent wildlife predators. The hardest job is the setup. The rest is a rewarding hobby or small farm business. Remember to recycle your chicken waste… it is wonderful fertilizer for your garden for vines planted on the outside of the covered chicken coop. I recommend allowing the large birds to free range on your property. They will eat your bugs and aging food thrown from the kitchen door. Be sure you throw the table scraps in the shade to lessen the chance of salmonella, shigella, and other germs that thrive on sugar and heat. How do these birds know your property lines? Perhaps they copy the dogs and cats. Our broody hens hatched their own eggs. One found out the hard way that stray cats and birds of prey will steal her young when she did not live under the mobile home and in the bordering shrubbery.

Pensacola, Bahia, or some other very small grass seed may be necessary to use the first week of life before the starter grower can be managed down the chicks' tiny throat and esophagus. The feed store had these seeds in supply in small, measured bags; do not get the ones treated with pesticides. Watch out for the roosters that come from a home raised as pets; they are mean and will fly at you and scratch if you are around their hens...particularly scary for a young child.

There is one big problem when living in a mobile home that does not have perfectly fitting skirting and without doors to block entrance of the hens from going under the mobile home. The hens like to roost where the insulation is, and will tear it away from your house and make a mess, not to mention lay dozens of eggs that may rot if you are not aware of their location. I detected the smell and discovered one nest when a rotten egg grazed my face from a pocket of insulation. There were so many of them that slight movement of the pouching insulation caused one of these old eggs to fall.

Would you really like to watch miraculous Mother Nature at work? Transplant small muscadine grapevines (or whatever is indigenous to your area) along the outside of your chicken coop and the covered fenced yard, as suggested earlier. Prepare to support heavily laden branches. The chicken manure will make your plants rapidly prosper, then apply yearly spring fruit set spray to facilitate an awesome grape production. I transplanted three Triumph (a variety of muscadine) vines in 1991 that yielded more than 120 pounds of grapes in 1995 and more than 220 pounds in 1996. And were they sweet...nature's candy! The entire pen had been completely shaded since 1995; a much-needed relief from the blistering Florida summer heat. As I mentioned, our grown flocks were free range, and they roamed from shade to shade by mid-morning. Besides ample shade, do not forget to set up the coop in a way that allows a good breeze. I found that crickets multiply rapidly under plywood. Turn over the garbage cans too so the chickens are able to get to them.

Our chickens and pullets (juvenile chickens) were like live garbage disposals and nothing went to waste. They loved slop and I have even tasted some after it was properly cooked. I was thinking of famine victims...not half bad, tasted like stew. The fowl also feasted on day-old bread and sweets from the local discount bread stores. The chickens roosted in a large Japanese plum tree (Loquat) in the front yard. Our roosters have been known to crow at any time, day or night, but they did not keep us awake and the neighbors never complained. One neighbor even had a herd of their own and these chickens occasionally visited our grounds, but did not mate with each other. Roosters will sometimes fight one another for a brief moment. A dead chicken has always been the result of a dog attack, except for two cases of fowl cholera. As I mentioned, ducks love water and excrete even in automatic bucket waterers. Unfortunately, puppies will chase fowl for sport, then when they are older to eat the fowl if not prevented. Eventually, the dogs get too old to want to try. Wouldn't it be nice to find a shepherd dog that protects rabbits and chickens and wouldn't dare chase them?! Do you think that these llamas that are rumored to protect sheep from wolves would do the same for rabbits and fowl? Until I actually see a film clip of this magnificent feat, I believe that the llama will run for his or her own life too if a wolf was nearby!

And do not forget to aid in the feeding of your ducks and geese at community ponds. A half rack of day-old bread sold for five dollars in 1994 and that will fill up a half of a small pickup truck bed. Sometimes the animal feed recipients that are on the stores' list will be called because there is a backlog of bread and/or sweets, and you may be offered a two-for-one deal. Watch ducks run toward you as they recognize your truck… that will indicate how hungry they are or not for the food.

Chapter Four

Dairy Calf Husbandry

Now, if you really want to help save the world, raise dairy calves. Their milk and manure are the backbone of survival for land dwellers. They may prove to be your toughest customers. It seems safe to say from my experience that if you do not have the time to baby-bottle feed them, they may be doomed to die. Sometimes they are already drinking via the bucket, but if they become ill, they may not feel like getting up and you have to bottle feed. We used the nine-ounce plastic human baby bottles and cut large holes in the nipples for our days-old bull calves. The commercial bottles for calves seem to be designed for older calves. If the ill calf (or calves) are not improving, it is time to take more drastic measures to ensure adequate fluid and electrolyte levels; and hence, survival. It is time to put a tube down the throat. I was once surprised to find that a dying calf was still interested in sucking, but when I went into the house for an electrolyte solution in order to correct the electrolyte imbalances from prolonged diarrhea, I found that he apparently went into cardiac arrest from hypokalemia (too-low levels of potassium) due to prolonged watery stools. The mineral blocks that can be purchased at feed stores may be enough, but I also needed the powered electrolytes per tube feedings when the calves experienced prolonged diarrhea and weakness. Unfortunately, I did not get to this little fellow in time. I should have started him on electrolyte water and mineral block as soon as I bought him from the dairy farm a few days earlier because he already had diarrhea. It can be very hectic on a farm, particularly when you are the only farm hand and raising a young child too. If the concentration of electrolyte water is too strong, it will burn the hair off the legs and whatever body part the watery stool touches. But that does not seem to hurt them. And I found that Kaopectate works wonderfully in alleviating excessive loose stools; our veterinarian had it in gallon jugs. I use to premix the milk and save it in the refrigerator, then reheated it for the next feed until I tasted how readily this milk soured. No wonder they had watery stools so easily. I wondered it was from the chickens getting in their feed buckets until I tasted the stored and freshly mixed milk. Do not feed homogenized/store-bought milk if you run out of their commercially prepared milk; that stuff really makes them sick.

I would suggest providing their milk (Suckle) for four to six weeks. I suspect stress fractures or complete fractures of the hip sometimes occurred either due to interbreeding, mineral deficiencies, or perhaps the mom accidentally laid on the offspring. The calf lies down for good. I have witnessed two out of twenty-four calves all of a sudden attempt to bear weight on their back legs, but they could not rise completely, and their hips fall to one side. They were standing when I bought them though, and even lasted a few days at my farm. I wish that I could have obtained hip X-rays of these two victims. Thank goodness for dog food manufacturers in our area; providing these calves to them is being resourceful.

The lovely dears were sometimes weak, but with no diarrhea, when I first got them from the dairy farm because they were probably only fed Suckle two times a day there and their solid food was not forced. I liked to feed them Suckle three times a day for the first two weeks and found that I had to force handfuls of sweet grain or day-old bread pieces into their mouths for the first few days. They ought to be eating pretty good by the end of the first two weeks. Mine seemed to prefer the torn-apart day-old bread pieces over grain, perhaps because it was softer. Always feed them two quarts of Suckle milk each meal. The smallest calves can be frustrating because they may not take the bucket when you are tired of bottle feeding; there is a product on the market with a nipple attached to the bucket that does help. One farmer suggested allowing the calf to suck the milk on your finger, then lowering it into the bucket of milk.

Do you need to find out how to do a tube feeding? I recommend acquiring a feeding bag with the balling tube attached to it from your large-animal veterinarian. Our veterinarian group happens to be a large and small-animal care provider, and I could not do without them; they are great. Anyway, just as Dr. Freel taught me, put the tube alongside of the corner of his or her mouth and allow him or her to swallow as you advance the tube. I used a landscape irrigation tube several times and checked placement with a 30-cc syringe and stethoscope before I found out there is a commercial product, but the little buggers chewed a hole in the line while instilling the milk and electrolyte water. Then you may really have problems should the little angels aspirate.

There may be a time when injectables (shots) are needed, such as antibiotics; they are much more effective than the oral ones. In nursing school, we were told that the previous class learned how to give them by injecting oranges, but we chose the option to practice on each others' thighs and arms and inject normal saline (sterilized salt water), which we chose. The calf's neck is the preferred site for intramuscular injections. Feed stores carry the necessary supplies. And, of course, there is the veterinarian. Thank God, there are some willing to go through extremely lengthy schooling to become veterinarians.

Our sweethearts were given two feedings a day of day old bread and/or grain. Sometimes their diet was supplemented with the aging fruit and vegetables from the grocery store. They even ate the icing and cakes from garbage refuse. They did not have the teeth and jaws yet for hard things like apples. The calves seem a little slow mentally as I mentioned, but they did eat well after a couple of weeks of forcing a handful of sweet grain in their mouths each morning for a few days. They could be quite hearty eaters of bits of not only day-old bread, but bananas, tomatoes, lettuce, etc. (soft fruit and small bite-size veggies) and sweet grain as well. To prevent the chickens from getting into their five-gallon buckets of grain, I suspended the buckets so they did not touch the ground. I put the calves in separate pins each Suckle feed until they were weaned between four and six weeks. Then they were turned out into a larger pasture with individual stalls to protect the grain and assist the young weanlings in reaching maturity.

There is the question of whether or not the bull calves need castration. Just as boars will bend down and crawl over fencing, so too will the bulls in their attempt to breed. I also learned the hard way that if a calf tries to fall down during castration and puts a lot of weight on his abdomen, your calf may have a hernia problem. His buddy next to him stood perfectly still and suffered no such dilemma, but Buttons had loose stool and probably was threatened by strangulation of his bowel. I opted for banding by the veterinarian for the later calves to facilitate castration, but I am not sure this is the best method for the animal.

You will never grow tired of looking at their beautiful, big black eyes and long black lashes. And I fear there are not enough small farmers to help the large dairy farmer. I was told by one helper at a dairy farm that the bull calves become alligator meat if no one will care for them. The dairy farmer does not have the space for them; they and the newborn heifers are kept in small pens just large enough to get up and to turn around. And there is little to no protection against rain, wind, cold, or extreme heat. But I understand there are calf hutches on the market that can be bought from a catalog and assembled (Nasco). I regret to say that we Americans are contributing to malnutrition worldwide by not ensuring that these calves reach maturity. I do not know the solution to this problem when these precious animals are merely dependents on a family, versus a cash cow, but I think the government ought to solve this problem and let every

Susan G. Mehaffey

bull calf and heifer be raised to full productivity. They do grow to enormous sizes! Plus we ought to have adequate shelters with our tax dollars to every young farm animal in the world. I think we owe it to them. And one of the greatest joys I have ever witnessed was the initial reaction of the bull calves to their new-found freedom on my farm. They may have landed "spread eagle" on all fours when they first took off running, but it did not take them long to get up and prance about. Watch how their little tails will wag! These poor babies on dairy farms…they really need our help.

Chapter Five

Organic Horticulture

Suppose that you live on drought-ridden, barren soil. Do you give up and sit on your "caboose"? No! Here is an excellent challenge. Create your own personalized fruit or nut orchard, or vineyard. Perhaps your project is sheer desertification, where you turn barren soil into enriched, composted soil. You will need a front-end loader to dig out two and a half feet deep pits, manual labor or backhoe to unearth every corner of the pits, and a dump truck or pickup to transport yards of dairy farm manure and composted soil from landfills or independent soil businesses. I manually shoveled into these pits alternating layers of dairy farm manure, sand (dug from the pits by the front end loader), leaves (from suburban driveways/properties), and the dark composted soil. I then mixed the medium well and repeated with four more layers of the same and mixed again. I may have put a layer of sawdust as well, but my notes do not reflect that. Basically you now have in-ground nursery pots. Then it is time to dig a little hole for your tree or vine in this mixed medium and transplant it. The plants will need good watering for a couple of weeks, and during droughts. We used a simple irrigation method; just move the hose over to the next pit once it is filled. The beauty of this is that there is no run-off of water; hence, no waste.

    Voila...you will simply be amazed at the growth rate! Do not forget to prune when needed and apply fruit set spray fertilizer every year when the blooms appear. Record your findings. Perhaps one day we will need to ask school children to recycle fruit and vegetable seeds from home and use the schools as distribution centers. If children are reading this text, underline or color code the sentences that point out the recycling ideas. Remember your state's agricultural newsletter when searching for, or willing to sell, agricultural products. It advertises plants, animals, equipment, and employment throughout the state. The miraculous growth may even allow fruit trees, vines, and shrubs to survive in cold, normally harsh climates for that kind of plant. I would not know myself. And I planted vegetables in the pits around the trees the first year…the plants were laden with two-pound tomatoes each and okra plants over six feet tall when standing at the edge of the pit.

    I am sorry that I cannot give you any more gardening tips; I never mastered this scientific art completely. I would have liked to experience rapid growth of an oak tree but never got around to this one. How exciting to watch the instant shade and nut production. I used manure more often than commercial fertilizer and ended up with far more weeds than prize yields in my gardens. I salute you, professional and amateur master gardeners. What a lot of tedious labor! Chickens do allow pesticide-free gardening, but they will cover your seedlings when they scratch the mulch looking for bugs. And if you feed your swine vegetables, you may find an abundance of sprouts in bare dirt spots. This must be how the South acquired so many musadine grapevines throughout forested areas. Refer to Chapter 3 about the proliferating viticulture due to chicken manure.

    I do want to stress the importance of that photo with my daughter and I next to two pitted peach trees and a small non-pitted peach tree in the center. I transplanted the five-foot tree in the middle of that picture in 1991 in a mere hole with no manure. I transplanted the other two in February 1992 and they were only five feet in height with three or four branches at the time of transplantations. I used the pit method described in the first paragraph of this chapter for the two trees on either side of us. I had fruit from these two trees in the summer of that very same year (1992). Granted, there was only one peach on three out of ten trees, but that seems rather remarkable to me. I never pruned any of the trees so my yield was not superb. I only encountered some white fungus on a few trees years later. They responded well to topical treatments with hydrogen peroxide. The trees and vineyard planted in the pits doubled in height in two years. I do not have much agricultural knowledge, but I think this might be ground-breaking news.

Susan G. Mehaffey

Diary of a Small Farmer

Susan G. Mehaffey

Diary of a Small Farmer

Susan G. Mehaffey

Diary of a Small Farmer

Susan G. Mehaffey

Diary of a Small Farmer

Susan G. Mehaffey

Diary of a Small Farmer

Susan G. Mehaffey

Diary of a Small Farmer

Susan G. Mehaffey

Diary of a Small Farmer

Susan G. Mehaffey

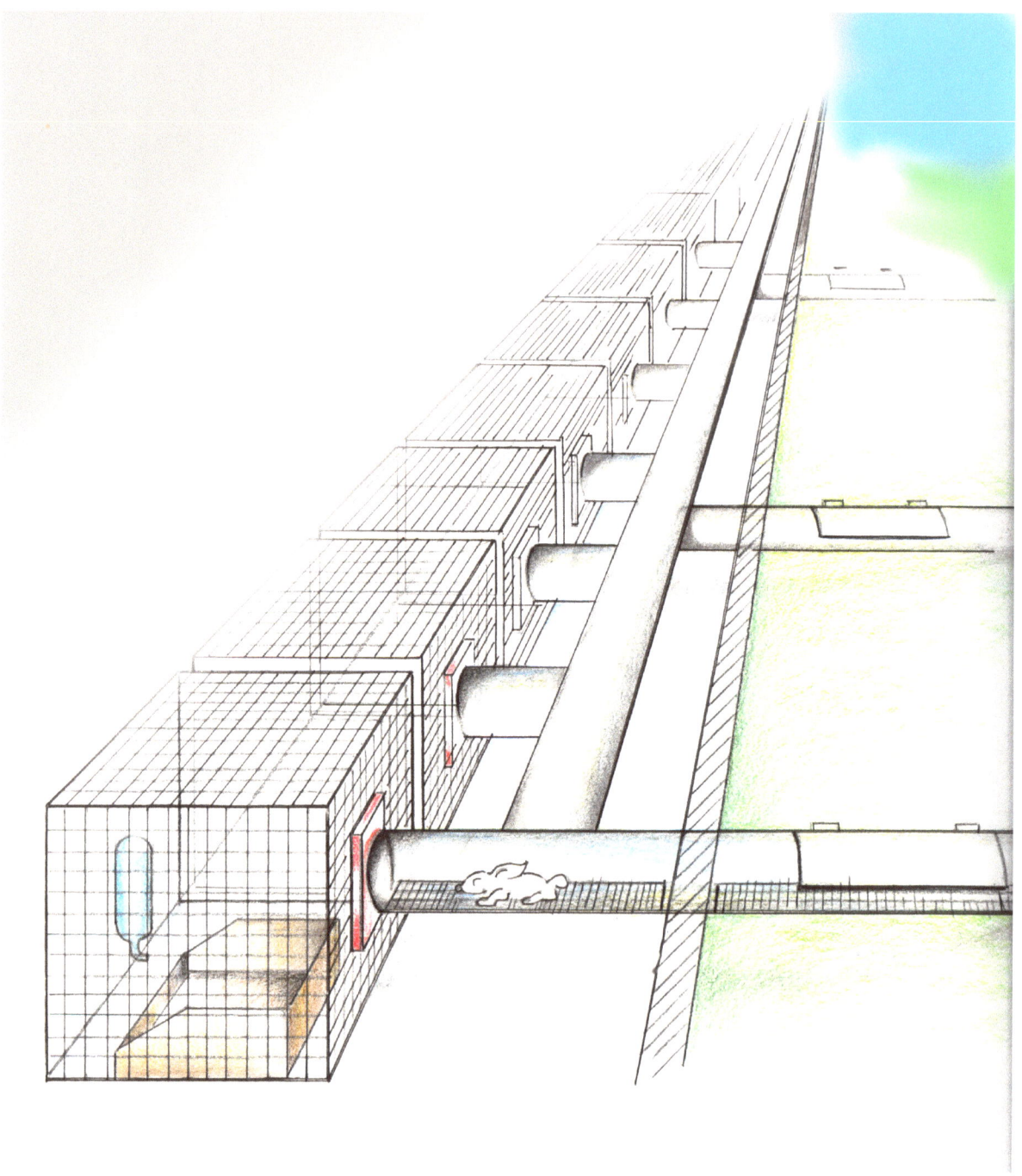

Diary of a Small Farmer

Susan G. Mehaffey

Epilogue

If you have never taken the opportunity to go to an annual 4H youth fair, I suggest that you see for yourselves how beautiful and well-fed all of the animals are, and some of them may be for sale for breeding purposes. Thank goodness for 4H students and their guardians. Perhaps there will one day be more 4H and FFA members in apartments and the suburbs than most can imagine at this time. And perhaps there will be an outcry from children and adults alike for free and volunteer bitubal ligation and vasectomy clinics in each city! I sure can appreciate that one! I dare say that sterilizations will be mandatory in some century. Perhaps it will be just the opposite, where we are all allowed to have children so that they will be used for seaburials one day. Personally I advocate for any woman to bear only one or two children at the most, no matter how many mates she might have in a lifetime. Let us not overpopulate the earth any quicker than impossible to prevent. Perhaps it does sound far-fetched, but is it inconceivable for shellfish and other sea life to feast upon dead carcasses and bodies at the bottom of the oceans?! Oceanography and marine biology research in needed whether or not is in a laboratory or in the field.

Let us work on efforts to prevent exploitation and premature deaths for the purposes of seaburials and only utilize accident or homicide victims and those who die of natural causes. Let us remember the words of Dr. John Frazer Hart, who writes on pages 375-376 of his book, *The Land That Feeds Us*, that "there will be no more land to encroach upon in the United States by 2500 [AD]." Let us start seaburial research and swill collecting now and be able meet the challenge of hungry victims of war, famine, and disease worldwide, not to mention be ready for 2500 A.D. Will adequate seaburial research inform us of a positive or negative environmental impact to the oceans?

I would like to advocate the refrigerated corpse display shown following this chapter in regards to seaburials. I am hoping to interest you in participating in the popularity of this new concept and its intended use. How resourceful are we being to future generations by simply burying in the ground or cremating? Are we poisoning our water, soil, earthworms, and other earthen organisms with embalming fluids? Humans, pets, and road kill are protein sources. Remember, it does take a lot to provide even one meal to our bodies so that it may turn into our necessary protein and energy. Let's think about human and/or pet seaburials seriously. Perhaps our greatest engineer, God, did have corpse and carcass seaburials in mind when He designed and built massive oceans with scavengers on the bottom and sharks with their powerful jaws and sharp teeth. I have watched a documentary on sharks feeding on dead whales; so apparently, their food need not be alive.

Let us not do away with embalming, but save it for such purposes as suspicious deaths. Let us preserve land for other things besides burial; needless to say we cannot grow more land. The refrigerated corpse display is simply a means to slow down the decomposition of our deceased loved ones and allow viewing at the funeral. The body can already be in a simple pine box used for transport and it can be recycled one day in the event of tree shortages. The refrigerated corpse display may be the transport of choice on airplanes or unrefrigerated boxcars for the deceased toward their final resting grounds in the sea. And it would save the millions of trees needed for the common coffins used today. Funeral homes relatively close to oceans can simply lead funeral processions toward the helicopter or boat transport without the refrigerated corpse display.

I envision hunting grounds and corralling of our shellfish one day. There will eventually be reefs of dead bodies extending from the north and south poles and around all the continents linking us all in unity. Let us remember to put out buoyant markers to designate our burial reefs, so that fishing nets do not cause the bodies to resurface. I imagine that cement blocks will be needed also to weigh down the corpses. We will have to send the helicopters carrying our precious cargo out as far as seven or more miles from our coastlines. Here is a suggestion as the simple rule: the burial reef could markt he borders of international waters. How far would that be now? I envision chartered funeral party boats and aircraft going to and from the seaburial reefs. And for those of you who worry about diseases being passed onto sea life: nursing school instructors made us realize why we irrigate dirty wounds with salt water or pack with saline dressings. The increased osmolarity of salt water causes swelling and bursting of the germ cells; hence, death of the disease.

So much to do and so little time to do it! Let us start today and not wait until crises arise. And if the crisis has arisen in your community, such as young animals being slaughtered, let us rise to the occasion and provide adequate quarters and care for them, so that they may reach their full potential and productivity. And let us not wait until scavengers of the sea become nearly extinct from centuries of routine "plucking." An old National Geographic magazine pointed out that professional fishermen are experiencing diminished catches.

I did do another amateur agricultural experiment that I would like to pass on. I had a freshwater ten-gallon tank and I fed ground beef and sirloin strips to crawfish and other bottom-feeding fish. I noticed they usually do not eat the sirloin strips probably because it is too hard. They seemed to eat some fat, since ground beef is marbled with it, but did leave a lot of fat that got caught up in the filter. This does make me wonder what would happen in nature…would the sea life make nest beds out of it or would it just be pollution?

What would happen to all the remaining bones? Perhaps there will be mining opportunities ahead after centuries of bone deposits. Do our motherless dairy calves need our bones to be harvested in order to supply adequate calcium and phosphorus in their commercialized milk? What other resourceful ideas come to your mind? Here are some unique thoughts: will we be driving in the ocean to take a break from crowds? Will we one day have underwater malls and submersible houseboats?! Will we one day be hunting and farming the fish and scavengers of the sea in these underwater vehicles? Will we sanction one-mile parcels of sea bottom for corralling our shellfish? Some of you might not be aware that some beekeepers keep hives in national forests on loaned parcels with a clean water supply. Perhaps these lessons could apply here too, to some degree. Will all retirees give up their land and homes to their children and live on the oceans? Perhaps the process of dehydrating fish by these young or old retirees will become popular one day and much can be stored before a land trip is necessary.

What amazing things can you invent? You can find the telephone numbers of an attorney and engineering firm (invention submission company) in the back of many magazines. I worked with Society of American Inventors based in Akron, Ohio. The reader may be interested in the picture on the next page. I wish to see a clear lid only on the RCD; the rendition by S.A.I. is their professional engineer's idea. I think it leaves room for smuggling. I would ask for a bottom line figure in writing. They did charge me over $9,000 and then I found out from the free information from the United States Patent and Trademark Office that a mere idea is not patentable!

Good luck in helping save the world and thank goodness for your caring, and do not forget to recycle…even yourselves if deemed environmentally sound. Perhaps we ought to ask university oceanography and marine biology departments to begin conferences as soon as possible and determine where to begin research. Would it be best for a laboratory simulated ocean environment or a natural setting, such as the small bay at St. Joseph State Park located along the Florida panhandle? For those of you who think only pets, zoo creatures, and the like should consume such farmed seafood, we ought to label it in grocery stores differently from the colonies not around sea burial reefs.

Do you want to know a good first way to get comfortable with the new image of becoming a farmer or farmers' helpers? Buy a pair of comfortable rubber boots that slip on and off easily;

try them on when you are overly tired…you will be grateful for this invention that takes little exertion. If you are a rancher or farmer, you might want to help your workers get a break from the heat by supplying an electric pole(s) with lamps for evening work. I had one installed and it made me feel more comfortable about late evening and night work. Onward mighty soldiers of Mother Nature and Father Time. And thank you for reading this simply to have an appreciation of what it takes to be a farmer or rancher and for the ideas mentioned. And for those of you who have a tendency to "wing it" year after year, be sure to plan far enough ahead to not suffer too much or set up your family or animals to suffer despite good intentions. Farming in hard work and it does take a lot of time, energy, and money!

Opinions do matter to our legislators and I imagine they are eager to do great things for the majority. A local representative, Cliff Stearns, seemed impressed after I introduced myself to him at a town meeting with a rough draft that I had sent him years ago. Let us contact our legislators and advocate the ideas presented in this book. First of all, let each of us begin to realize that the remaining large parcels of undeveloped land in the Deep South need to be preserved for agribusinesses and their support services. Let us take proper measures to ensure a healthy future for all tomorrow no matter how congested humans must exist one day.

There is another subject I would like to include in this book. I do want you to be aware of a very serious occupational hazard around forests. Some ticks carry the parasite ehrlichhia, which can cause ehrlichiosis. This disease can leave you gasping for air and antibiotics are a necessity to prevent convulsions and death. Ferrous sulfate tablets (iron) will help arrest the acute respiratory distress syndrome, but that medicine will not make the attack of air-hunger disappear altogether and the iron tablets will eventually stop working. I saved myself with Kelfex that my father had on hand. It did take 2500 mg at one time; the 500 mg tablets every six hours just did not do the trick. There were no repercussions after taking so much.

I would like to close this book with these timeless quotes found on the walls of the Presidents' Hall at EPCOT, a part of the Disney World empire:

> Thomas Wolfe: "I think the true discovery of America is before us. I think the true fulfillment of our spirit, of our people, of our mighty and immortal land is yet to come."

> Charles Augustus Lindbergh: "What kind of man would live where there is no daring? I don't believe in taking foolish chances. Nothing can be accomplished without taking any chance at all."

REFRIGERATED CORPSE DISPLAY

Environmental Harmony...

KEEP LOVED ONES LOOKING GOOD THROUGH THE ENTIRE VIEWING CEREMONY.

COOLING SYSTEM BUILT INTO BOTTOM OF CASKET SENDS FRESH AIR UP THROUGH VENTS IN THE FORM-FITTED BODY MOLD.

THIS REUSABLE VIEWING COFFIN IS ENCASED HALF IN GLASS WITH A SOLID LID FOR CLOSED CASKET CIRCUMSTANCES.

Manufacturing/Distribution Information: 1-888-USA-IDEA
email: invent@inventorshelp.com (1-888-872-4332)

Patent Pending   Registered Professional Engineer's Feasibility Statement Available   Reg. US PTO

Summary

Here is the summary of the ideas throughout the book and their purposes. Perhaps they will become a politician's platform one day:

1) <u>Free-range habitats</u> so as to allow unsurpassed quality of life for small farm animals and farmers. Perhaps other humans' aesthetic need can be met too, such as providing a facility for the dying or disabled to view the activities.

2) <u>Force-fed organic pits</u> with dairy farm manure so to facilitate rapid growth and production.

3) Insure all young <u>farm animals grow to full productivity</u> so to prevent malnutrition and hunger in the world.

4) Insure <u>adequate quality of life for all farm animals</u> so to prevent suffering of these precious resources.

5) <u>Human and euthanized pet seaburials</u> without the use of embalming fluids in order to feed the shellfish and other marine life after proper long-term research, if deemed environmentally friendly

6) <u>Swill collection centers</u> for recycling of table scraps in order to have ready-made supplies for present and future farmers and perhaps even for victims of famine.

7) <u>Fenced food carts for farmers</u> near grocery bins, preventing wastefulness of aging food so to supplement the diets of caged and pastured farm or zoo animals. Quite frankly, I ate a little of this food too after properly washed and cooked... here is one bonus if you are a poor farmer or wish to attempt farming on a small budget.

8) <u>Bitubal ligation and vasectomy clinics in every city</u> offering services based on voluntary admissions in order for population growth control and prevention of unwanted pregnancies.

9) <u>School-aged children recycling seeds</u> from home in order for prevention of vitamin and mineral deficiencies relieved by fruits and vegetables.

10) <u>Houseboats that can convert to sub-oceanic vehicles</u> for work and play because of the now more than 6 billion people in the world and growing, plus hurricane seasons to endure.

About the Author

Susan Goulard Mehaffey is a divorced mother of one, retired RN, retired farmer, first-time author, and innovative thinker of new ideas.

www.ingramcontent.com/pod-product-compliance
Lightning Source LLC
Chambersburg PA
CBHW051046180526
45172CB00002B/533